SMALL BUT DANGEROUS

Written by Paul Stevenson

CONTENTS

Small but Dangerous	4
Invisible Killer	6
Bullet Ants	8
Poison Puffer	10
The Deathstalker	12
Stonefish	14
Blue Rings for Danger	16
Piranhas	18
Killer Frog	20
Poison Arrows	22
Deadly Snails	24
Deadly Spider	26
World's Biggest Killer	28
A Deadly Bean	30
Glossary	31
Index	32

First published in 2024 by
Hungry Tomato Ltd
F15, Old Bakery Studios,
Blewetts Wharf, Malpas Road,
Truro, Cornwall,
TR1 1QH, UK.

Copyright © 2024 Hungry Tomato Ltd

No part of this publication may be reproduced, stored in a retrieval system, or transmitted in any form or by any means, electronic, mechanical, photocopying, recording, or otherwise, without prior written permission of the copyright owner.

A CIP catalogue record for this book is available from the British Library.

ISBN 9781916598782
Printed in China

Discover more at
www.hungrytomato.com

Neither the publisher nor the author shall be liable for any bodily harm or damage to property whatsoever that may be caused or sustained as a result of conducting any of the activities featured in this book.

All words in **BOLD** can be found in the glossary.

Small but DANGEROUS!

What would you prefer to meet: a snarling tiger or a delicate butterfly?

Large, fierce animals can be dangerous.
However, some of the biggest killers on Earth...

...are smaller than a coin!

The tsetse fly, from Africa, lives by sucking blood from people and animals. As it does this, it spreads a disease.

The disease makes people fall asleep, and never wake up!
This killer fly kills many hundreds of people each year.

THE TSETSE FLY MAY BE SMALL, BUT IT IS VERY DANGEROUS!

INVISIBLE KILLER

The most venomous animal on Earth is the box jellyfish. It lives in the Pacific Ocean.

Its tentacles have millions of dart-shaped stingers. When it fires the stingers into its **victims**, it releases **venom**.

A sting from this invisible killer can kill a person in less than 4 minutes.

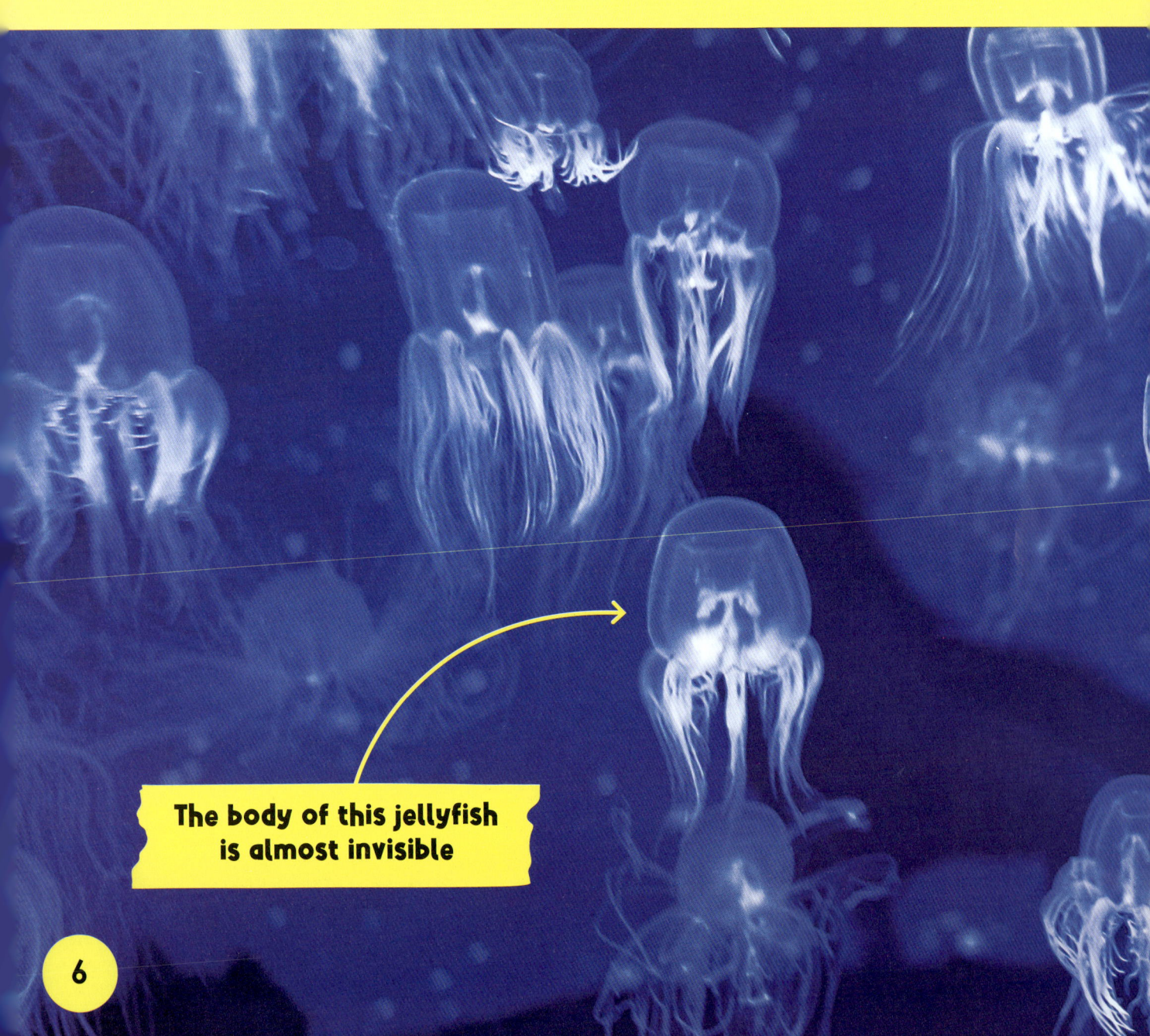

The body of this jellyfish is almost invisible

VINEGAR IS A LIFESAVER!

Vinegar stops the stings from working. They can then be removed.

A vinegar station on a beach for treating jellyfish stings

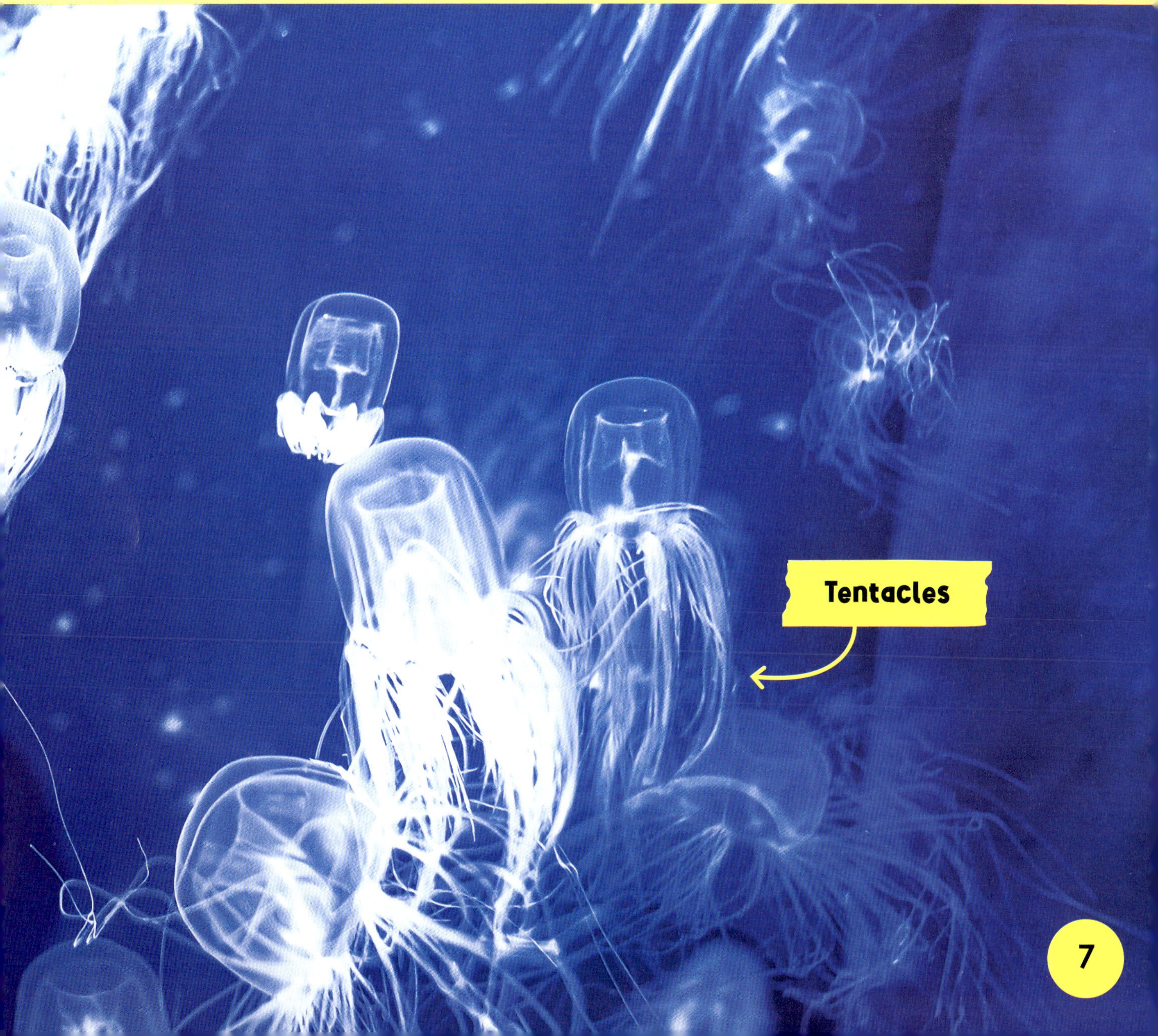

Tentacles

BULLET ANTS

What's the most painful insect sting on the planet? The common wasp or bee? Not even close!

According to bullet ant victims, its sting feels like being shot! That's how it got its name.

One person described the pain in more detail:

> "It's like fire-walking over flaming charcoal with an 8cm rusty nail in your heel."

Stinger

Bullet ant

When boys of the Sateré-Mawé tribe in Brazil reach 13 years old, they collect bullet ants. The ants are weaved into a pair of gloves, which the boy must wear for a full 10 minutes without screaming.

POISON PUFFER

The pufferfish makes a strong, deadly poison inside its body.

The poison is stored in its liver.

If a person eats poisonous pufferfish flesh, it paralyses their muscles. This causes them to stop breathing and suffocate to death!

Pufferfish in danger

When in danger, pufferfish puff up into a round ball. This makes it hard for a **predator** to bite them.

THE DEATHSTALKER

The deathstalker is one of the deadliest scorpions in the world. It is found in the deserts and scrubland of the Middle East and North America.

Its tail is full of powerful venom, which can paralyse its victim.

The front claws are used to catch prey

This venom, however, is also very valuable to humans. It's used in medicine to treat cancer and **malaria**, and has been called "the most expensive liquid in the world".

THE DEATHSTALKER HAS THE HIGHEST AMOUNT OF VENOM OF ALL SCORPIONS. IT'S VERY DANGEROUS!

The venom comes from the stinger in the scorpion's tail

STONEFISH

The stonefish is one of the most venomous fish on Earth! It lurks amongst rocks on **coral reefs**.

If you step on a stonefish, the spikes on its back jab into your foot. The spikes can even jab you through a shoe.

Venom then squirts into your skin. This venom makes it hard for you to breathe.

YOUR FOOT WILL BE IN AGONY!

WHAT TO DO IF YOU STEP ON ONE:
- Put your foot in hot/warm water to help the pain.
- Seek medical attention right away to get treatment.

BLUE RINGS FOR DANGER

The blue-ringed octopus grows no bigger than a golf ball, but it's still deadly!

They only show their stunning blue rings when alarmed.

Bites from blue-ringed octopuses are rare, but these animals carry venom that's **lethal** to humans.

The venom will paralyse you and prevent you from breathing.

YOU WILL BE DEAD IN MINUTES.

Blue-ringed octopus

In Thailand, people buying octopuses in markets are told to report anybody selling blue-ringed octopuses. There is no **antidote,** and the venom is deadly even after cooking!

PIRANHAS

Piranhas are little fish with big teeth. They live in rivers in South America.

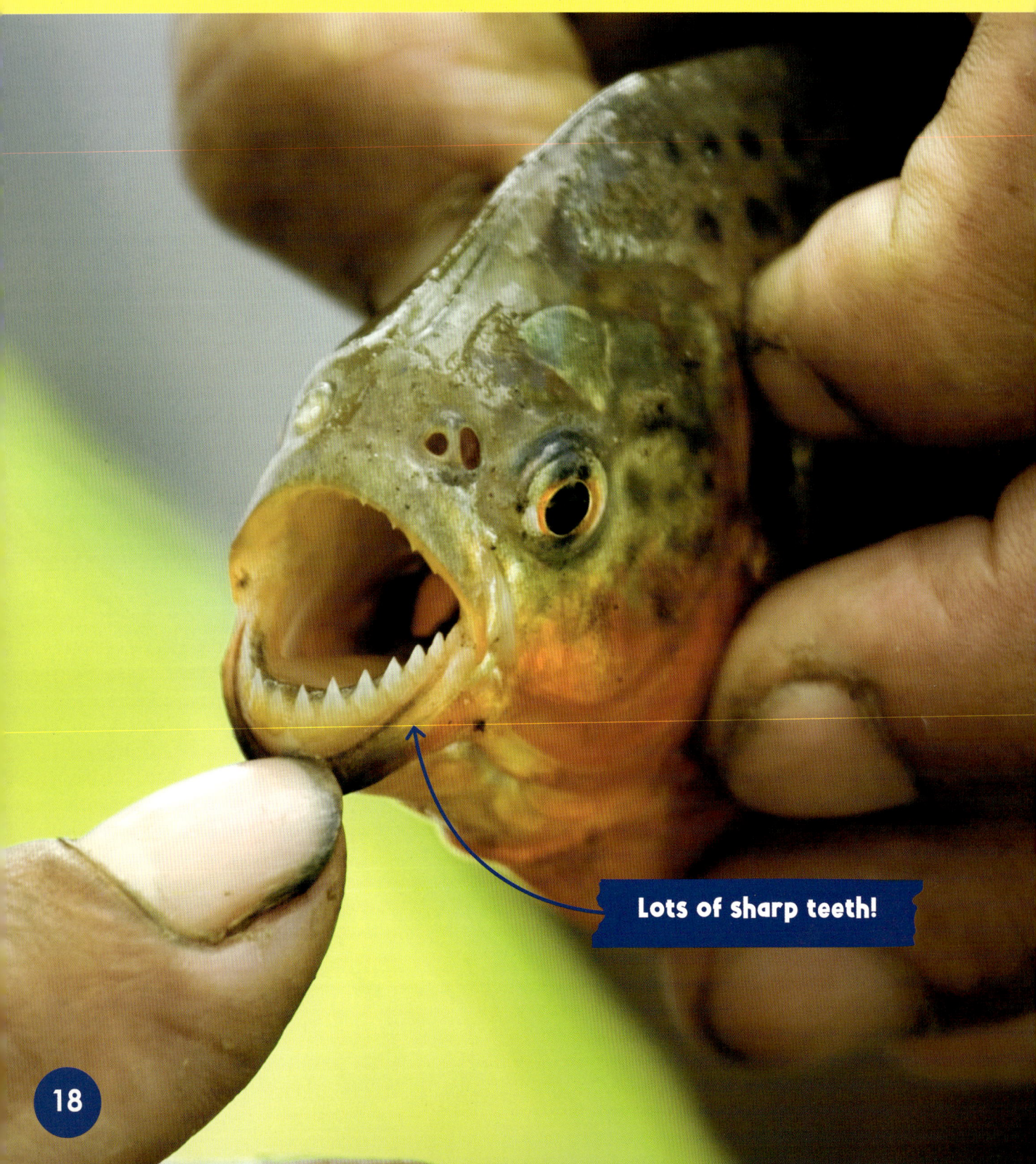

Lots of sharp teeth!

One piranha can give a nasty bite, but if piranhas gang together they can be DEADLY!

Sometimes, rivers start to dry up when it's hot. Lots of piranhas have to live together in a small amount of water.

The piranhas then hunt as a pack.

When a large animal, such as a cow, comes to drink – **the piranhas attack!**

The victim is stripped to its bones!

KILLER FROG

The golden poison frog is less than 5cm long. However, some scientists think it might have enough poison to kill 10 people!

The poison comes from the ants and beetles that the frog eats.

Just touching the frog gives you a painful rash.

If a predator bites the frog, the predator will be dead in minutes!

The golden poison frog lives in the **rainforests** of Colombia in South America.

The poison works through the frog's skin

21

POISON ARROWS

Hunters in the rainforest use poison from the golden frog on their arrow tips.

Any animal pierced by the arrow will be paralysed in seconds!

The arrow tip is rubbed on the frog's skin

DEADLY SNAILS

Scientists sometimes have dangerous jobs, including milking snails! These aren't your average snails; they are deadly cone shells that live in the sea.

Scientists collect venom to make new medicine, such as painkillers.

Cone shells have a **harpoon**. They use it to spear fish.

Breathing tube

Harpoon comes from here

Cone shells pump venom into their victims through the harpoon.

THEY PUMP ENOUGH VENOM TO KILL 15 PEOPLE!

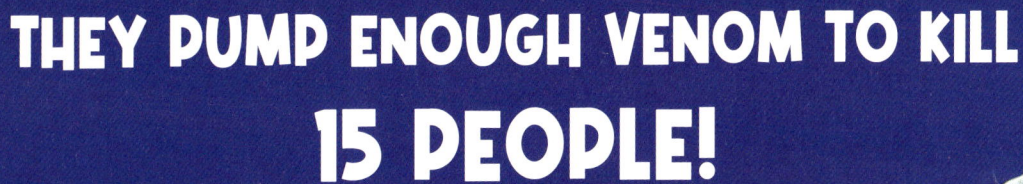

Harpoon seen through a microscope

Bad news:
- No antidote exists.

Good news:
- The known number of humans killed is less than 100.

It's best to avoid touching shells in tropical waters, in case they are alive!

DEADLY SPIDER

The Sydney funnel-web spider is one of the world's most venomous spiders. It lives in Australia.

The spider has a deadly bite; its venom will make you sweat and vomit. It will also make you twitch - you won't be able to control it.

Within a few hours, you may die from a swollen brain.

HOWEVER, THERE IS AN ANTIDOTE TO THE VENOM. SO GET TO A HOSPITAL FAST!

A scientist milks the spider for its venom so it can be used to make an antidote

These spiders often fall into swimming pools. They can survive underwater for up to 24 hours.

Funnel-web spiders are known for spinning their webs into a funnel shape

WORLD'S BIGGEST KILLER

The most dangerous animal on Earth is the tiny mosquito. Every year, millions of people catch diseases from mosquito bites.

Mosquitoes live by sucking blood from people and animals

Mosquito repellent

When mosquitoes bite people, they spread deadly diseases, such as malaria and **yellow fever**. Nearly every minute, a child under 5 dies of malaria.

A DEADLY BEAN

One of the deadliest living things is not an animal...

...but a BEAN!

The castor bean contains a deadly poison called ricin. If pure ricin gets into your blood, it can kill you. A pellet of ricin the size of a pinhead can kill a person in hours.

In 1978, Russian spies used ricin to kill a man called Georgi Markov. Using a mini airgun hidden inside an umbrella, the spies fired a tiny pellet, filled with ricin, into Markov's leg.

Castor bean plant pods with beans growing inside

GLOSSARY

agony - a terrible pain.

antidote - a drug that stops a venom or poison from killing a person.

coral reef - a rocky area in warm, shallow seas. A coral reef is made of the chalky remains of huge numbers of tiny animals called coral polyps.

deadly - something that causes death.

harpoon - a type of arrow or spear.

lethal - something that causes death.

malaria - a disease that causes liver and brain damage. It is most common in Africa and Southeast Asia.

paralyse - made unable to move all or part of the body.

poison - a substance that can kill or hurt a person or animal.

predator - an animal that hunts and kills other animals for food.

rainforest - a jungle of tall trees that grows in hot, wet parts of the world.

suffocate - to die because you can't breathe.

venom - a poison that is deliberately passed onto a victim through a bite or sting.

venomous - an animal that uses venom to kill prey or to defend itself.

victim - a person or animal who is hurt or killed.

yellow fever - a deadly disease that causes liver and kidney damage. It is mostly found in South America and Africa.

INDEX

A
Africa 5, 31
antidotes 17, 25, 26, 31
arrows 22-23
Asia 31
Australia 26

B
bites 17, 19, 26, 28-29
blue-ringed octopus 16-17
box jellyfish 6-7
bullet ants 8-9

C
castor beans 30
Colombia 20
cone shells 24-25
coral reefs 14-15, 31

D
deathstalker 12-13
disease 5, 28-29, 31

G
golden poison frogs 20-21, 22-23

H
harpoon 24-25, 31

M
malaria 13, 29, 31
Markov, Georgi 30
medicine 12-13, 24
mosquitoes 28-29

P
Pacific Ocean 6
paralysis 10, 12, 17, 22, 31
piranhas 18-19
poison 10, 20-21, 22, 30-31
pufferfish 10-11

R
rainforests 20, 22, 31
ricin 30

S
scorpion 12-13
South America 18, 20, 31
stings 6-7, 8-9, 12-13
stonefish 14-15
suffocation 10, 31
Sydney funnel-web spiders 26-27

T
tentacles 6-7
Thailand 17
tsetse flies 5

V
venom 6, 12-13, 14, 17, 24-25, 26-27, 31
vinegar 7

Y
yellow fever 29, 31

Picture credits:
(t=top; b=bottom; c=centre; l=left; r=right):
Shutterstock: 14-15, 19tl, 19b, 30; Adam Calaitzis 7tr; Andrei Antipov 8c; Elizaveta Galitckaia 29tr; Evannovostro 31b; frank60 28-29; FtLaud 10c; KRIACHKO OLEKSII 4; Kurit afshen 20-21; Nuttawut Uttamaharad 6-7b; shaftinaction 1, 27.
A.N.T Photo Library/NHPA: 24. Birgitte Wilms/Minden Pictures/FLPA: 16. Istock: 2-3, 10-11, 18. Jeffrey L. Rohman/Corbis: 17c. John Chapple/Rex features: 26b. Jurgen Freund/Nature Picture Library: 9. Mark Moffett/Minden Pictures/FLPA: 22-23. Martin Dohrn/Science Photo Library: 5c. Volker Steger/Science Photo Library: 25.

Every effort has been made to trace the copyright holders, and we apologise in advance for any unintentional omissions. We would be pleased to insert the appropriate acknowledgements in any subsequent edition of this publication.